Human Emotions

BY CARRIE SMITH

Table of Contents

Introduction

You feel them every day. A car speeds through a puddle on a rainy day and splashes you with muddy water. You're angry. You get an invitation to a friend's party. You're happy and excited. You see a hungry, stray dog on your way to school. You feel sad. **Emotions**, whether strong or weak, are always with you. You may not always like them, but it's hard to get rid of them. Have you ever thought about just what an emotion is?

Emotions are complicated. They begin in the brain and work in ways that scientists are only now beginning to observe and understand.

This book will explain the important role that emotions play in your life and how they affect the body and the brain. It will also show you how to understand and manage your emotions in positive ways.

What Is an Emotion?

Most people think of an emotion as a **feeling**. But that description may not be accurate. If you touch a hot stove burner, you "feel" pain, but that is not an emotion. If you touch the hot stove burner, feel pain, and get angry with yourself for doing such a foolish thing, then you are experiencing an emotion.

Scientists who study human emotions believe that people could not survive without them. Emotions are part of the human species' built-in survival kit. Emotions make you react to things that happen in your environment. They can get you to act in some way or not act at all. They can often help keep you safe.

Imagine that you are crossing the street. Suddenly, a car appears. It is coming directly toward you—fast. You feel fear. Your mind seems more alert than it was the instant before. Your body seems more alive, too.

You don't have time to think about what to do. You just do it. You get out of the car's path in the nick of time. Your emotion—in the form of fear—has prevented you from getting hurt.

The fact that people from different countries share the same facial expressions allows them to recognize one another's emotions.

THINK IT OVER!

The word *emotion* comes from the Latin verb *motere*, which means "to move." Have you ever heard someone say that they felt "moved" by something? If so, they were talking about having an emotional reaction.

No matter where you live, what language you speak, or what your background is, you feel the same emotions as every other human on the planet.

Scientists have identified six emotions that all people seem to show in the same way. These six **universal emotions** are happiness, sadness, fear, anger, surprise, and disgust. For example, a surprised person in the United States will raise his or her eyebrows in the same way as a surprised person in Asia.

The fact that people from different countries share the same facial expressions allows them to recognize one another's emotions. It means that even though you may not share the same language, you can understand and respond to one another.

happiness

sadness

fear

anger

surprise

disgust

In addition to the six universal emotions, there are other emotions that scientists refer to as **social emotions**. Unlike universal emotions, social emotions are not always shown by facial expressions. As a result, they are often harder to recognize.

Scientists believe that people learn social emotions. Guilt is one example of a social emotion. A newborn baby does not feel guilt. In order to feel guilt, you need to know that you have done something that you shouldn't have and be able to feel bad about it. Babies can't think in those ways.

There are many social emotions. In addition to guilt, there are jealousy, embarrassment, shame, love, hate, and grief.

How many different emotions do you feel in any given day? Think back to yesterday. What happened? Make a list of all of the different emotions you felt and why you felt them.

▲ Just like older people, babies can feel sad, happy, angry, surprised, disgusted, and afraid. But lacking language and experiences, they can't yet feel some of the social emotions.

THE STUFF OF GREAT STORIES

The social emotions are the building blocks of many great stories. Writers use them all the time. The playwright William Shakespeare was a master at this. In *Romeo and Juliet*, he used the two emotions of love and hate to create a timeless tragedy. Two families hate each other, but their children have secretly fallen in love.

▲ William Shakespeare (1564–1616)

How Emotions Form

Sometimes you feel emotions because of something that you experience through one or more of your senses—seeing, hearing, tasting, touching, or smelling. These are called **sensory experiences**. Here are some examples: Your uncle hands you two box seat tickets to a major league play-off game. One look at them and you're jumping up and down with joy.

You come home from school in a bad mood because you have a big report to write. You turn on the radio and your favorite song is playing. Suddenly, you feel a lot better.

You're with your family driving along a local highway. You pass several factories whose smokestacks are spewing fumes into the air. The smell is strong and unpleasant. You feel disgusted by the smell and angry at the people who pollute the air.

Emotions sometimes form because you think, imagine, or remember something.

Emotions aren't always sensory experiences. Emotions sometimes form because you think, imagine, or remember something. In other words, your mind can spark an emotion all by itself. For example: You're climbing a mountain. You're doing just fine until you think, "What if I lose my footing and fall?" A picture forms in your mind in which you are plunging off the cliff. You start sweating and your heart starts racing. You lean into the rock and all your muscles tighten. Suddenly you're frozen with fear.

You're taking a walk and you find yourself thinking about your grandmother, who died last year. You remember how the two of you used to take walks together. You realize how much you miss her. Before you know it, you're feeling sad. Your eyes start to tear.

You get home before your parents do. You're alone in the house for a few minutes. There's something in the basement you want to get, but as soon as you think about going there alone, you get scared. Although you know there's no one down there, you can't help thinking, "What if there is?" You listen for noises. You start to feel nervous. You decide to wait on the porch until someone else gets home.

?

Do you make associations with any foods? Which foods? What are the mental connections you make with each?

Sometimes emotions are caused by mental connections called **associations**. Certain foods can often cause associations because you connect the foods with another time in your life. Some of the associations you make with food may be pleasant. Others may not. It often depends on whether you like the food or not.

EMOTIONS AT THE MOVIES

Did a movie ever scare you? Alfred Hitchcock was a famous film director who really knew how to scare people. How did he do it? He started by using main characters an audience could identify with. Then he made those everyday characters feel trapped or chased by people or things they couldn't see or control. His characters usually had nowhere to go for help or protection. They had to solve a problem on their own.

Hitchcock's settings were very real. And he knew exactly what to show on camera and what to leave to the audience's imagination. By forcing his audience to use their own imagination, he made them bring their own fears right into the movie theater.

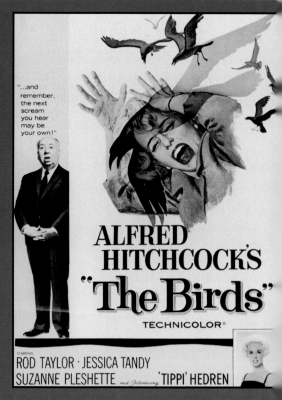

"...and remember, the next scream you hear may be your own!"

ALFRED HITCHCOCK'S "The Birds"
TECHNICOLOR®

STARRING ROD TAYLOR · JESSICA TANDY
SUZANNE PLESHETTE and Introducing 'TIPPI' HEDREN

THINK IT OVER!

Remember a time when you felt a strong emotion such as happiness, anger, or embarrassment. Can you make yourself feel that emotion again just by replaying the memory in your head?

Do you remember your first birthday cake? Your first hot dog at a baseball game? Your first plate of spinach? A Thanksgiving turkey may remind you of a very pleasant holiday spent with family. Eating popcorn may make you remember your first time at the circus. Which of these foods do you like? Which don't you like? What foods make you especially happy?

How Emotions Change the Body and the Brain

Emotions cause many changes in the human body. You probably already know about some of these changes. The most obvious ones affect your skin, heart, muscles, and blood vessels.

The muscles of your face shape your mouth into a smile or a frown. Your skin blushes with embarrassment or turns pale with fear. Your palms get sweaty when you are nervous. Your heart races with excitement. Tears roll down your cheeks when you are sad. These are some of the obvious changes that your emotions can cause. Other changes happen inside your brain, but they can be observed only with the help of new technologies.

pale with fear

heart pounding

smiling

tears rolling down face

blushing

frowning

Part of the brain translates your sensory experiences, memories, and mental pictures into emotions.

Scientists have discovered that most emotions are formed in a certain part of the brain. This part of the brain translates your sensory experiences, memories, and mental pictures into emotions. Then it sends messages to other parts of the body that cause you to react to the emotions. If this part of your brain didn't work, you would not be able to experience emotions.

Many emotions form in the part of the brain below the **cerebral cortex**. This part of the brain is active in different ways during different emotions. ▶

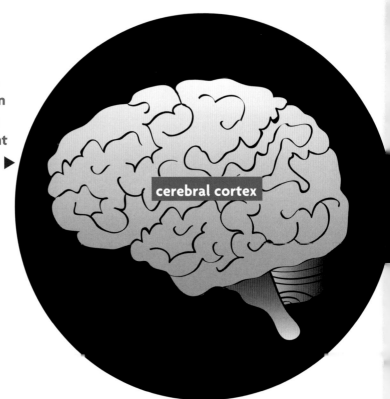

cerebral cortex

ADRENALINE

Everyone feels fear at some time. When was the last time you were afraid? How did your body react? When you feel fear, your heart rate increases, you breathe faster, and your senses are sharper.

These reactions are caused by a chemical called **adrenaline** (uh-DREH-nuh-lin). When you feel fear, your brain sends a message to your **adrenal glands**. Then your adrenal glands, which are located on top of your kidneys, release adrenaline into your bloodstream.

Do you know why this happens? It is to help you. The adrenaline prepares you to face a difficult situation. It gives you the strength and the focus you need to either fight the danger or flee from it as fast as you can.

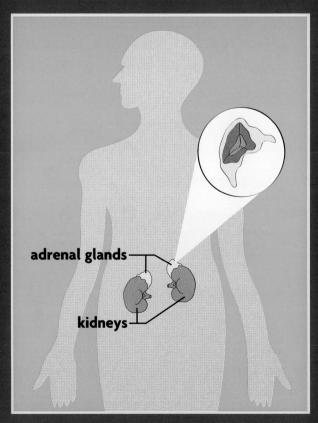

adrenal glands

kidneys

Comparing Human and Animal Emotions

Do you believe that animals have emotions? Millions of pet owners do. They don't need any more proof than what they observe every day. Are you familiar with any of the following situations?

You see a dog wagging her tail furiously, and you say, "Look, the dog's happy!"

Another dog bares his teeth at you and growls, and you think, "I better not get any closer. That dog is angry."

You bring home a new kitten. Suddenly your ten-year-old cat won't sit on your lap anymore. He hisses and arches his back. You decide, "He's hurt—and jealous."

There is no question that many people have special relationships with their pets. They seem to be able to understand each other and **communicate** with each other. Some people even treat their pets as they would treat a person—dressing them, making special foods for them, talking to them as if they could understand every word. They do this, in part, because they believe that their pets have the same kinds of emotions that people do.

Do you think animals have emotions? Why or why not? Compare your ideas with those of a friend or classmate.

Many researchers who study animals in the wild believe that animals experience universal emotions such as fear, anger, and happiness.

Scientists have many different opinions. Some believe that all animal behavior is driven by **instinct** only.

Many researchers who study animals in the wild, however, believe that animals experience universal emotions such as fear, anger, and happiness. Some of these researchers even think that certain animals feel social emotions such as love and grief.

They base their thinking on what they have observed: dolphins chasing each other through the water playfully; chimpanzees dancing and somersaulting like children; elephants appearing to mourn deeply for dying or dead family members. Elephants have even been known to stay by the body of a dead animal for days, touching it with their trunks and trying to wake it up.

▼ Dolphins swim and jump playfully.

▲ Chimpanzees play, fight, and show concern for one another.

▲ Elephants form long-lasting relationships with family members and miss them when they are gone.

The part of the brain where emotions form is similar in animals and humans.

Researchers have discovered that emotions form in the brains of some animals the same way they do in the brains of humans. In fact, the part of the brain where emotions form is similar in animals and humans.

Animals react to emotions in similar ways, too. For example, fear causes an animal's heart to beat faster and makes its senses sharper. Fear helps animals fight or flee from danger the same way it helps humans.

▲ A turtle hides in its shell when it is scared.

▲ A porcupine extends its quills for protection.

Many scientists believe that the big difference between human and animal emotions has to do with awareness. Humans are **self-aware**. They can think about their emotions while they are having them. Most animals don't seem to be able to do this. Although they may have emotions, they may not be aware that they are having them.

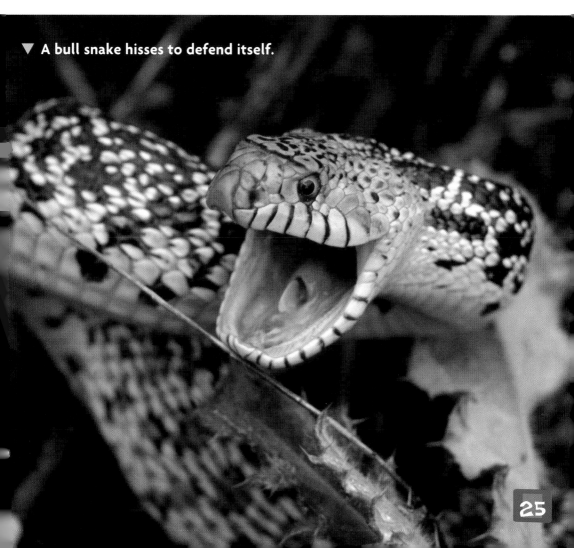

▼ A bull snake hisses to defend itself.

The Importance of Knowing Your Emotions

"I can do this."

Being self-aware means that you can pay attention to your emotions as you have them. In this way, you can learn from your emotions. For example, you can begin to see how certain situations trigger certain emotions in you. Knowing ahead of time how you may feel in a certain situation gives you the opportunity to avoid the situation or prepare for it. You can often control your reactions if you are self-aware.

Here are some examples:

Three minutes before a class presentation, your heart is pounding. From past experience, you know that if you get too nervous, you're not going to be able to speak. So you start taking deep breaths and letting them out slowly. You tell yourself, "I can do this." You're still nervous when the teacher calls your name, but you do a great job.

"I forgot my bib."

You're having a fight with a friend and you really want to say something hurtful. But you don't. Despite your anger, you value the friendship and don't want to risk ruining it.

You spill something on your shirt in the lunchroom. When you get to your next class, everyone stares at the stain. You feel embarrassed. You wish you could hide. But instead, you smile and say, "I forgot my bib." Everyone laughs—including you— and your embarrassment goes away.

CONSCIOUS VS. UNCONSCIOUS

Sigmund Freud (ZEEK-munt FROYD) was an Austrian doctor and professor who increased people's understanding of self-awareness. Freud's study of human behavior led him to believe that people often act on emotions even when they are not aware, or conscious, that they have them. He believed that these hidden emotions get revealed through people's dreams or through things they do and say. The term "Freudian slip" refers to a slip of the tongue that reveals something someone didn't mean to express—an emotion buried in the unconscious mind.

Successful people, no matter what they do, make their emotions work for them, not against them.

Being aware of your emotions is part of what makes you human. It enriches your life and can give you the motivation to do many positive things. All successful people, no matter what they do, make their emotions work for them, not against them.

Artists and writers use their emotions to create amazing works of art. Scientists pay attention to their "gut feelings" as they search for answers. Protesters use their anger to change society. Explorers conquer their fears in order to go to uncharted places. Athletes use the disappointment of defeat to help them achieve victories.

ACTORS AND EMOTIONS

Many of the world's greatest stage and screen actors use their own emotions to help them show emotion in a performance. When a role demands that they express a strong emotion, they look for an experience in their own past that made them feel something similar. They use this experience to create real emotion while acting. This is called Method acting.

CHAPTER 6

Using Your Emotions

ou can use your emotions in positive ways, too. Here are just a few examples:

• You find an abandoned kitten that does not look healthy. It hurts you to see it suffer. You put the kitten in a box and take it to a shelter. Two days later, it gets adopted.

• A manufacturer is polluting a river near your home. You have gone canoeing on that river and know how beautiful it is. You get angry and write letters to the local newspaper, your state representatives, and environmental organizations. They all start pressuring the manufacturer to stop polluting the water.

• You're extremely sad over the death of a good friend. You write a poem to read at the memorial service. After the reading, people tell you how much your poem meant to them. You get to share your grief with others, and you feel a little better.

As you can see, emotions can help you do many positive things. Can you think of an emotion you have had that you could use in a positive way?

Glossary

adrenal glands   (uh-DREE-nul GLANDZ) two small parts of the body, located on top of the kidneys, that release adrenaline

adrenaline   (uh-DREH-nuh-lin) a chemical in the body that is released by the adrenal glands when a person experiences fear

associations   (uh-soh-shee-AY-shunz) mental connections between objects and emotions

cerebral cortex   (seh-REE-brul KOR-teks) the outer layer of the brain

communicate   (kuh-MYOO-nih-kate) let someone else know something

emotion   (ih-MOH-shun) a strong feeling such as happiness, sadness, fear, or anger

feeling   (FEE-ling) a physical sensation experienced through one of the five senses

instinct   (IN-stingkt) the tendency to act or react to something on impulse, without thinking

self-aware   (self-uh-WAIR) able to pay attention to one's emotions

sensory experiences   (SEN-suh-ree ik-SPEER-ee-en-ses) emotions experienced through one or more of a person's senses—seeing, hearing, tasting, touching, and smelling

social emotions   (SOH-shul ih-MOH-shunz) learned emotions, including jealousy, embarrassment, guilt, shame, love, hate, and grief; they are harder to recognize than universal emotions because they are not always shown by facial expressions

universal emotions   (yoo-nih-VER-sul ih-MOH-shunz) the six emotions all humans experience—happiness, sadness, fear, anger, surprise, and disgust

Index